International Science

Workbook 2

International ▌▌▌▌▌▌▌ Science

Workbook 2

Karen Morrison

HODDER
EDUCATION
AN HACHETTE UK COMPANY

Hachette UK's policy is to use papers that are natural, renewable and recyclable products and made from wood grown in sustainable forests. The logging and manufacturing processes are expected to conform to the environmental regulations of the country of origin.

Orders: please contact Bookpoint Ltd, 130 Milton Park, Abingdon, Oxon OX14 4SB. Telephone: (44) 01235 827720. Fax: (44) 01235 400454. Lines are open 9.00–5.00, Monday to Saturday, with a 24-hour message answering service. Visit our website at www.hoddereducation.co.uk.

Cover photo © Lester Lefkowitz/Corbis
Illustrations by Robert Hichens Designs and Macmillan Publishing Solutions
Typeset in 12.5/15.5pt Garamond by Macmillan Publishing Solutions
Printed and bound by CPI Group (UK) Ltd, Croydon, CR0 4YY

A catalogue record for this title is available from the British Library

ISBN 978 0 340 96607 5

Contents

Measuring in science

Activity 1 **Converting from one unit to another**   Date:_____

When you want to change a measurement to a larger unit than the one you have, you need to divide.

When you want to change a measurement to a smaller unit than the one you have, you need to multiply.

The amount by which you multiply or divide is decided by how many places you are moving. The diagram below shows you how this works for different units.

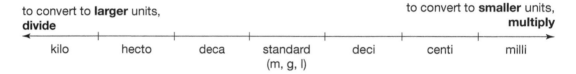

In the diagram, each place represents a power of 10. So you can see that when you change a measurement from millimetres to metres, for example, you have to divide by 1000 (10^3), because you are moving three places to a larger unit.

You can also see that if you change from kilometres to metres you have to multiply by 1000 (10^3), because you are moving three places to a smaller unit.

Working with square units and cubic units

When you want to convert square units, such as cm^2 to mm^2, or m^2 to hm^2, you have to remember that each place represents 10^2, or 100.

When you want to convert cubic units, such as cm^3 to mm^3, or m^3 to hm^3, you have to remember that each place represents 10^3, or 1000.

To convert square or cubic units, you apply the same rules:

- when you change from a larger unit to a smaller unit, you multiply (by 100 or 1000) for each place you move

- when you change from a smaller unit to a larger unit, you divide (by 100 or 1000) for each place you move.

1 Convert each of these measurements to the units given.

a) 24 cm = _____ mm

b) 250 g = _____ kg

c) 390 km = _____ m

d) 7500 m = _____ km

e) 45.6 ml = _____ cl

f) 59 cm = _____ m

2 Complete this table to show what quantity you could measure with each measuring instrument, and what units you would use most often in science.

Measuring instrument	What you can measure with it	Units used most often in science
stopwatch		
vernier ruler		
speedometer on a car		
weighing scales		
burette		
callipers		
spring balance		
measuring cylinder		

Activity 2 **Estimating and measuring length**

Date:_____

1 Estimate (guess) the length of each item in the table. Estimate in millimetres if the item is less than 1 m long, or in metres if the item is longer than 1 m. Write your estimate in the correct column.

Item to be measured	Estimated measurement	Measurement in millimetres	Measurement in metres
length of one brick			
width of one brick			
height of one brick			
diameter of a cool drink can			
width of a ruler			
length of classroom wall			
height of classroom door			
width of classroom door			
length of teacher's table			
length of pupil's desk or table			
length of this book			
thickness of this book			
your own height			
length of your foot			

2 Use a metre stick or tape measure to measure the length of each item in millimetres or metres. Write the result in the correct column.

3 Convert all the millimetre measurements to metres, and all the metre measurements to millimetres.

Activity 3 — Reading scales

Date:_____

1 Write down the amount of liquid shown in each measuring cylinder.

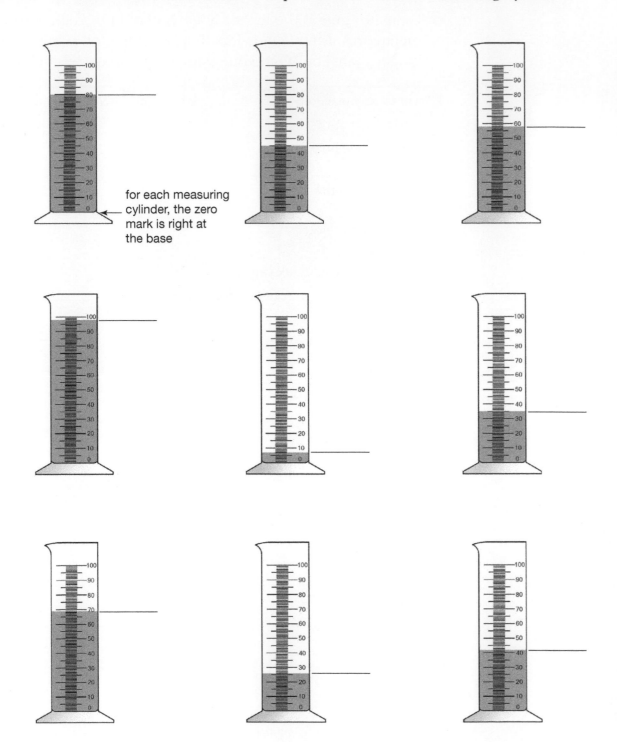

for each measuring cylinder, the zero mark is right at the base

2 On each measuring cylinder, draw a line to show where the liquid will rise to when a stone of volume 12 cm³ is dropped into it.

 **Measuring mass**   Date:_____

1 Estimate the mass of each item in the table, in grams or kilograms. Write your estimate in the correct column.

2 Use a kitchen scale or a beam balance to measure the mass of each item. Write the mass in the correct column.

Item to be measured	Estimated mass	Measured mass
pencil		
this book		
a litre of water		
a small stone		
a tube of glue		
a wristwatch		
a cell phone		
a packet of crisps		
one shoe		
one shoe-lace		
a comb		
a pencil case full of items		
a brick		
an apple		

Activity 5 **Units of time** Date:_____

1 Complete this table by working out how many seconds and how many minutes are in each unit of time.

Unit	Symbol	Number of seconds	Number of minutes
millisecond			
second			
minute			
hour			
day			
week			
month (31 days)			
month (30 days)			
year			
leap year			

2 a) Work out how many hours you have been alive.

_____

_____

_____

b) How many minutes is this? _____

c) How many seconds is it? _____

3 Label this diagram of a stopwatch.

Food and digestion

Activity 1 **Drawing a balanced meal**   Date:_____

1 Divide this plate into sections to represent the different food groups and the proportions in which you should eat them. Label each section.

2 Cut out pictures of food or draw food to show different things that you like to eat from each group.

Activity 2 **The digestive system** Date:_____

1 This diagram shows the human digestive system.
Use the diagram to complete the table.

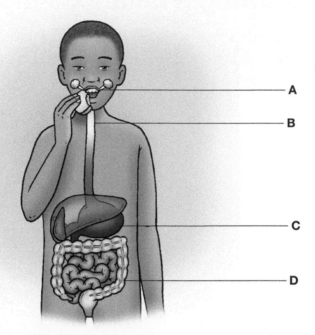

Part	Name of part	Function of part
A		
B		
C		
D		

2 Draw lines to join the bun, hamburger and cheese to the food group
they belong to.

carbohydrates

fat

protein

3 What could you eat with this hamburger to have a healthier meal?

_____

_____

Activity 3 **Describing digestion**

Date:_____

Copy and complete this flow diagram to summarise the parts of the
human digestive system.

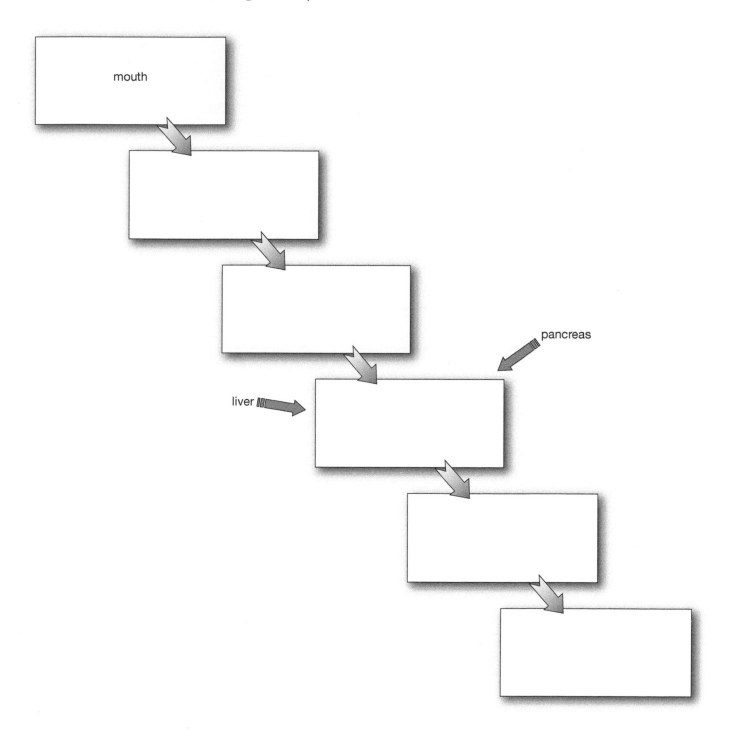

Activity 4 Working with statistics

Date:_____

If you eat more food than your body needs, your body stores the extra food in the form of fat and you become overweight.

People who are very overweight are classified as **obese**. **Obesity** is a growing problem in many countries and health experts are concerned. It is particularly worrying that obesity among very young children is increasing. This is happening because many children do not exercise a great deal, and have diets that are high in fat and sugar.

The table shows you the percentages of men and women who are classified as obese in different regions of one developing country.

Region	A	B	C	D	E	F	G	H	I	whole country
Men	10.1	8.1	10.2	10.4	6.2	7.5	7.6	5.5	13.1	9.3
Women	29.7	29.2	35.6	35.4	20.1	25.8	24.8	18.9	31.2	30.1

1 Look at the table. Write down five pieces of information that it tells you about obesity in this country.

_____

_____

_____

_____

_____

2 Draw a combined bar graph to show the obesity levels for men and women in each region, compared to the national levels (the country as a whole).

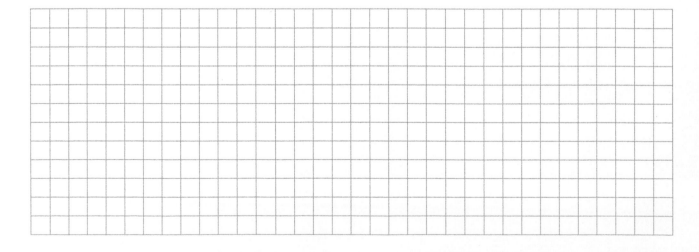

3 Study the weight-for-height charts below.

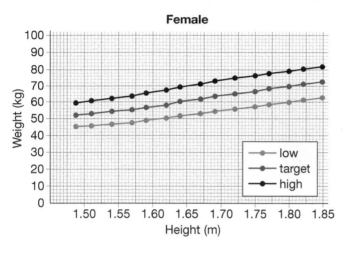

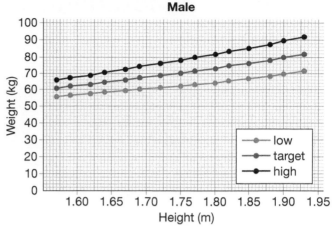

a) What is the normal weight range for a young woman who is 1.5 m tall?

_____

b) How would you classify a young man who is 1.6 m tall and who weighs 75 kg?

_____

c) What are the recommended weight limits for young men between 1.6 m and 1.7 m tall?

_____

_____

Activity 5 **Observing feeding adaptations**

Date:_____

1 Match the birds in the drawings to the descriptions of beaks below. Write the name of each bird under the correct description.

* fine pointed beak for catching and eating small insects

   _____

* strong hooked beak for tearing the meat off the bones of its prey

   _____

* strong curved beak for pulling insects out of holes in the ground

   _____

* strong straight beak for spearing fish

   _____

* strong cone-shaped beak for cracking open seeds

   _____

* long thin curved beak for reaching nectar inside flowers

   _____

2 On separate paper, draw the beaks of two other birds and show how they are adapted to the foods they eat.

Circulation and breathing

Describing blood Date:_____

1 Colour this pie graph, and complete the key, to show what percentage of human blood is made of each component. Each section of the pie represents 5%.

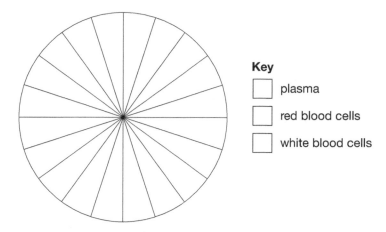

Key

☐ plasma

☐ red blood cells

☐ white blood cells

2 Why do you think the pie graph does not include platelets?

_____

3 This diagram shows what blood looks like through a microscope.

   a) Colour the red blood cells red.

   b) Colour the white blood cells purple.

   c) Colour the platelets green.

   d) Colour the plasma pale yellow.

   e) Count the white blood cells.

   How many are there? _____

   f) Estimate how many red blood cells there are. How many do you estimate?

   _____

Activity 2 · Counting heartbeats

Date:_____

1 You can feel your heartbeat as a pulse.
The pictures show you how to feel your pulse.

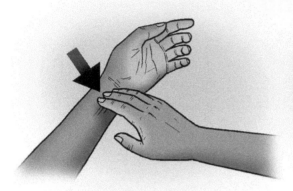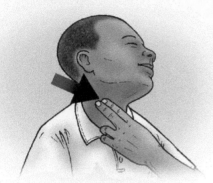

Take your pulse with a finger, not your thumb. Your thumb has a
pulse of its own, which could confuse your measurements.

a) Count your pulse for 15 seconds: _____ beats

b) Work out how many times your heart beats in one minute:

_____ × 4 = _____ beats per minute

c) Complete this table to show how
many times your heart beats in
1 minute, 1 hour and 1 day.

Time	Number of heartbeats
1 minute	
1 hour	
1 day	

2 Look at how many times your heart beats in 1 minute. Hold a tennis
ball in your hand. Try to squeeze the ball that many times.

a) How long did it take you to do this? _____ minutes

b) How did your hand and arm feel afterwards?

_____

c) What does this tell you about your heart muscle?

_____

3 Run on the spot for 1 minute.

a) Count your pulse for 15 seconds again: _____ beats

b) Write this as beats per minute: _____ beats per minute

c) What has happened to your pulse rate?

_____

d) Why do you think this happened to your pulse?

_____

Activity 3 Labelling diagrams

Date:_____

1 Give the diagram below a title.

2 Label each part on the diagram with its correct name.

_____

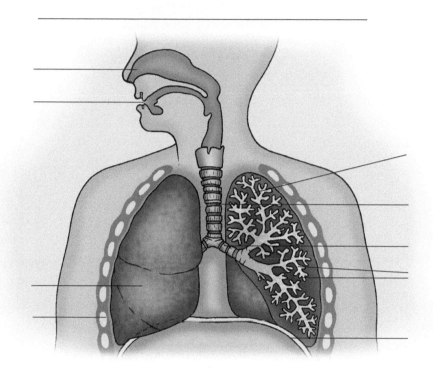

3 Label the diagram of alveoli and a capillary to show how gas exchange takes place in the lungs. Use the labels from the box.

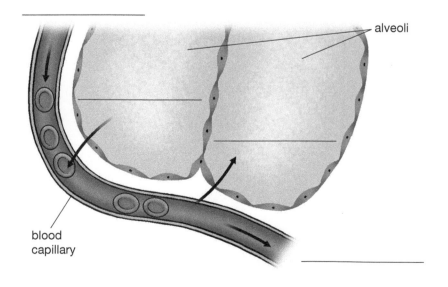

alveoli

blood capillary

deoxygenated blood

carbon dioxide moving by diffusion

oxygenated blood

oxygen moving by diffusion

Activity 4 **Testing air for carbon dioxide**

Date:_____

You can test a gas to see if it contains carbon dioxide by bubbling it through limewater. If there is carbon dioxide in the gas, the limewater will turn milky or cloudy.

Set up two test tubes of limewater as shown in this diagram.

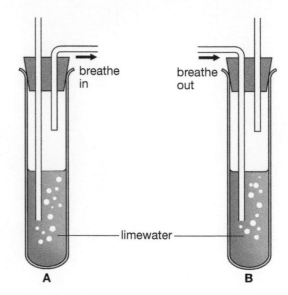

Breathe in through the bent tube in test tube A. This will cause air to move into the straight tube from outside the test tube, through the limewater, and into your mouth.

Breathe out through the bent tube in test tube B. This will cause the air from your lungs to pass through the limewater and out of the straight tube.

Continue inhaling through A and exhaling through B.

1 What happens to the limewater in test tube A?

_____

2 What happens to the limewater in test tube B?

_____

3 What does this tell you about the air you breathe in and the air you breathe out?

_____

_____

Activity 5 — Comparing lung capacity

Date:_____

You are going to do an experiment to find out how much air the lungs of young people, older people, smokers and non-smokers can hold. The amount of air that the lungs can hold is called the **lung capacity**.

Aim

To measure and compare the lung capacity of various people.

Apparatus

- ten balloons (all the same shape and size)
- a large basin of water
- a ruler

Method

Find ten people to do your experiment.

Ask each person to breathe in, and then to blow out into a balloon, with one exhalation.

Close the neck of the balloon.

Measure the volume of air in the balloon by putting it under water. Using your ruler, find the difference between the water level with the balloon (B) and the water level without the balloon (A). This difference (the value of $B - A$) represents the volume of air in the balloon.

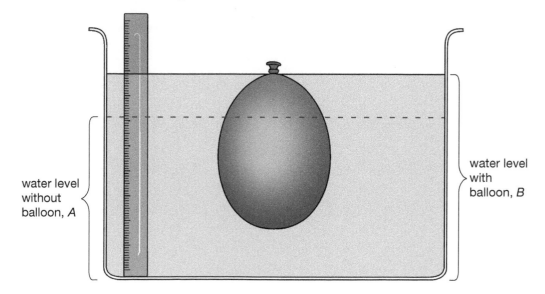

water level without balloon, A

water level with balloon, B

Record your results in this table.

Name of person	Age	Smoker (yes or no)	B – A

Questions

1 Does lung capacity increase or decrease as you get older?

_____

_____

2 Do fit people have greater or smaller lung capacity?

_____

_____

3 How does the lung capacity of smokers compare with that of non-smokers?

_____

Suggest a reason for any difference.

_____

Respiration

Activity 1 ## Comparing inhaled and exhaled air

Date:_____

1 Colour and label these two pie graphs, and complete the key, to show the differences between inhaled and exhaled air.

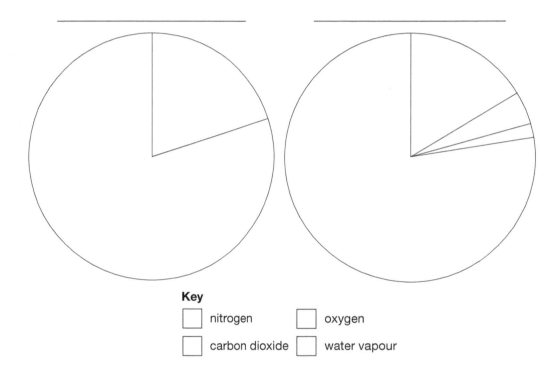

Key

☐ nitrogen   ☐ oxygen

☐ carbon dioxide   ☐ water vapour

2 Explain, using an equation, why there is less oxygen in exhaled air and more carbon dioxide in exhaled air.

_____

_____

_____

_____

Activity 2 **Interpreting graphs** Date:_____

The graph shows the amount of lactic acid in the blood of an athlete who exercised vigorously for 10 minutes.

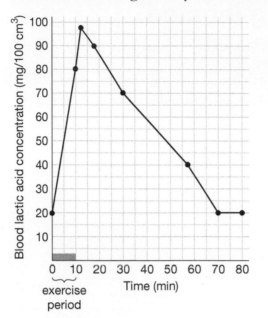

Time (minutes)	Lactic acid in blood (mg/100 cm³)
0	
10	
20	
30	
40	
50	
60	
70	
80	

1 Use the graph to find the information to complete the table.

2 How much lactic acid was in the athlete's blood before the exercise?

_____

3 Describe the change in the lactic acid concentration as shown on the graph.

_____

_____

4 Explain why there was a sharp increase in lactic acid concentration during exercise.

_____

_____

5 What causes the concentration of lactic acid to drop after exercise?

_____

_____

6 How long does it take for the lactic acid concentration to return to the level it was at before the athlete started exercising?

_____

Activity 3 Wordsearch

Date:_____

Twelve key terms to do with respiration are hidden in the wordsearch grid.

Find the key terms and list them below.

Then write a short definition for each word.

O	E	F	O	O	D	C	G	L	O	C	D	W
X	R	N	X	Y	A	L	A	H	X	S	E	A
Y	P	G	G	E	N	E	R	G	Y	G	B	T
G	Y	C	W	T	A	X	A	L	G	L	D	E
E	X	H	A	L	E	D	C	A	E	U	I	R
N	L	R	E	D	R	O	I	T	N	C	O	B
D	U	B	R	O	O	N	B	C	O	O	X	T
E	A	T	O	X	B	T	H	I	F	S	Y	K
B	W	A	B	Y	I	N	H	A	L	E	D	P
T	E	R	I	G	C	D	E	B	O	N	E	M
Q	L	A	C	T	I	C	A	C	I	D	B	L
C	A	R	B	O	N	D	I	O	X	I	D	E

1 _____

2 _____

3 _____

4 _____

5 _____

6 _____

7 _____

8 _____

9 _____

10 _____

11 _____

12 _____

Activity 1 **Drawing plant root systems** Date:_____

Draw the correct root systems on these plants.

- Plant A has a tap root system. The main root is about the same length as the plant's height.

- Plant B has a fibrous root system. The roots go down to a depth of about a quarter of the plant's height.

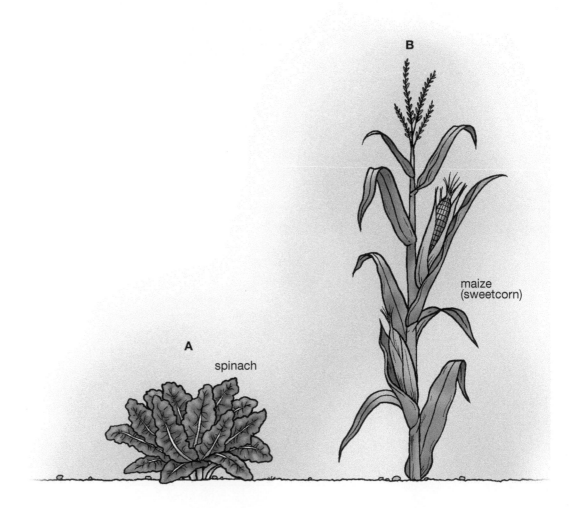

B

maize
(sweetcorn)

A

spinach

Activity 2 Measuring plant roots

Date:_____

The diagram shows how deep into the soil different plant roots grow.

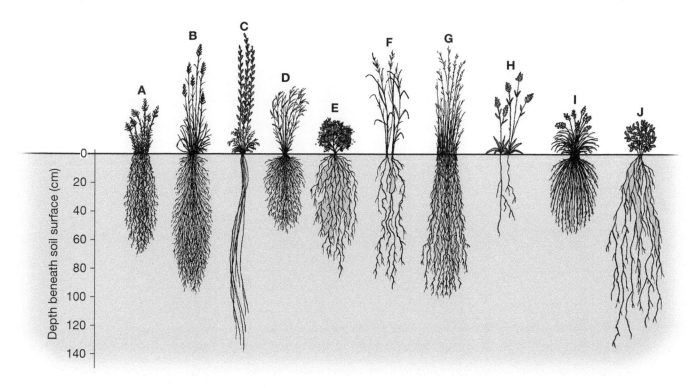

1 Complete this table to show the length of each plant's root system.

Plant	Length of root system (cm)
A	
B	
C	
D	
E	
F	
G	
H	
I	
J	

a) Which *two* plants have the longest roots? _____ and _____

b) Which *two* plants have the shortest roots? _____ and _____

2 Draw a bar graph on the grid below to compare the lengths of the two longest root systems and the two shortest root systems.

- Use a scale of 1 cm per 10 cm on the vertical axis.
- Arrange the bars in height order from longest to shortest roots.
- Give your graph a title.

Activity 3 Labelling a diagram of a plant

Date:_____

Label each part of the plant in the diagram with its correct name.

Activity 4 · Comparing flowers

Date:_____

1 Look at these flowers from eight different plants.

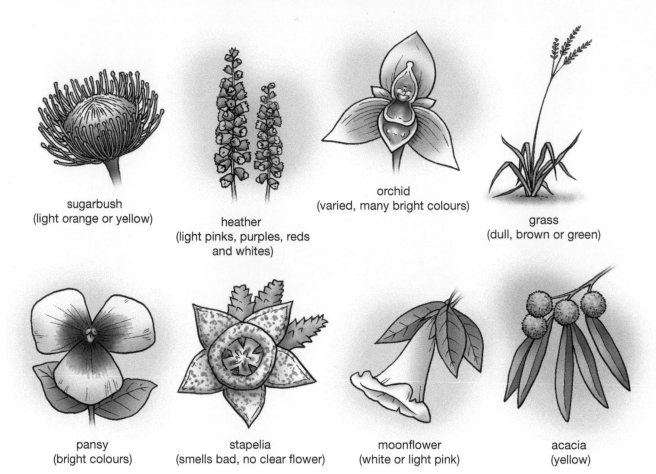

sugarbush
(light orange or yellow)

heather
(light pinks, purples, reds
and whites)

orchid
(varied, many bright colours)

grass
(dull, brown or green)

pansy
(bright colours)

stapelia
(smells bad, no clear flower)

moonflower
(white or light pink)

acacia
(yellow)

2 Complete this table to describe the flowers found on each plant.

Plant	Colour	Shape of petals	Number of petals
sugarbush			
heather			
orchid			
grass			
pansy			
stapelia			
moonflower			
acacia			

Activity 5 **Test your knowledge**   Date:_____

Write one scientific word that matches each of these descriptions.

1 The type of bud that develops at the node of a plant.
_____

2 The reproductive part of a plant. _____

3 The area of a stem between any two nodes. _____

4 A small stalk that attaches a leaf to a plant. _____

5 The structure that supports the flower. _____

6 The plant structure that anchors the plant and absorbs water from the soil. _____

7 A side offshoot of the stem of a plant. _____

8 The main support of the plant. _____

9 The part of the plant that produces food from sunlight.
_____

10 Where a leaf develops on a stem. _____

11 Thick, main root of a plant. _____

12 The bud at the tip of the stem. _____

13 Tubes that transport water in a plant. _____

14 The part of a flower that grows into the seed. _____

15 A system of tubes for transporting food in plants. _____

16 Develops in the seeds of a plant. _____

Atoms and elements

Elements in our bodies

Date:_____

The table shows you the percentages of different elements found in our bodies.

Element	Percentage of human body made of each element (%)
hydrogen	63
oxygen	25.5
carbon	9.5
nitrogen	1.4
calcium	0.31
phosphorus	0.22
chlorine	0.03
potassium	0.06
sulphur	0.05
sodium	0.03
magnesium	0.01
all others	<0.01

Use the information from the table to complete the pie graph and key below.

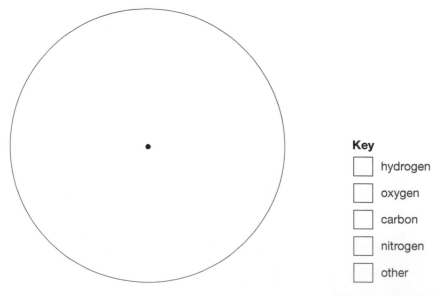

Key

☐ hydrogen

☐ oxygen

☐ carbon

☐ nitrogen

☐ other

**Activity 2** **Elements found in the Earth's crust**

Date:_____

The bar graph shows you the percentages of different elements found in the Earth's crust.

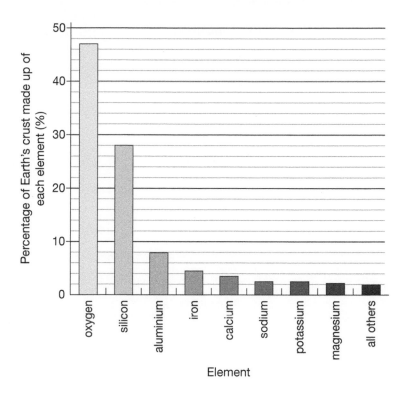

Use the information from the graph to complete this table.

Element	Percentage of the Earth's crust (%)

Activity 3 # Identifying atoms

Date:_____

Study the diagrams of different atoms carefully.

Write the atomic number of each atom.

Use the Periodic Table on page 56 of your coursebook to find out the name of each element.

2 protons

atomic number:_____

name of element:_____

8 protons

atomic number:_____

name of element:_____

12 protons

atomic number:_____

name of element:_____

11 protons

atomic number:_____

name of element:_____

17 protons

atomic number:_____

name of element:_____

Activity 4   Working with the Periodic Table

Date: _____

1 Fill in the chemical symbols and missing atomic numbers on this Periodic Table.

hydrogen								helium
H								2 **He**

lithium	beryllium		boron	carbon	nitrogen	oxygen	fluorine	neon
Li	4		**B**	6 **B**...	**N**	8	**F**	**Ne**

lithium	beryllium	boron	carbon	nitrogen	oxygen	fluorine	neon
Li	4	**B**	6	**N**	8	**F**	**Ne**
sodium	magnesium	aluminium 13	silicon **Si**	phosphorus 15	sulphur **S**	chlorine **Cl**	argon 18
Na							
potassium	calcium 20	gallium 31 **Ga**	germanium 32 **Ge**	arsenic 33 **As**	selenium 34 **Se**	bromine 35 **Br**	krypton 36 **Kr**
K							
rubidium 37 **Rb**	strontium 38 **Sr**	indium 49 **In**	tin 50 **Sn**	antimony 51 **Sb**	tellurium 52 **Te**	iodine 53 **I**	xenon 54 **Xe**

2 Shade the metals green, the metalloids yellow and the non-metals blue.

3 Circle the elements in this list.

water	sugar
air	salt
vinegar	oxygen
brass	krypton
calcium	neon
bread	zinc
silver	carbon

Activity 5   Talking about atoms and elements

Date:_____

Use what you know, and the glossary on page 155 to 158 of your coursebook, to help you write a definition for each of these scientific terms used in this chapter.

Term	Definition
matter	
atom	
element	
compound	
nucleus	
proton	
atomic number	
neutron	
mass number	
electron	
Periodic Table	
chemical symbol	

Chapter 7

Molecules and compounds

Activity 1 **Writing formulae**

Date:_____

Write the formula for each of these molecules.

a) N—N

b) H—O—H

c) P P P P (ring)

d) H C H with H top and H bottom

e) F—F

f) O O O

g) Cl H—C—H Cl

h) H—Cl

i) F Xe F

j) N with H H H

k) H C N

l) H H H C O H

m) H H—C—C—H with H's

n) Cl Cl Cl S Cl Cl Cl

o) H—Br

p) H O O H

q) I I

r) S ring (S₈)

a) _____ g) _____ m) _____

b) _____ h) _____ n) _____

c) _____ i) _____ o) _____

d) _____ j) _____ p) _____

e) _____ k) _____ q) _____

f) _____ l) _____ r) _____

Activity 2 Identifying parts of compounds

Date:_____

1 Complete this table by filling in the names of the elements that make up each compound and the number of atoms of each element. The first example has been completed for you.

Compound	Formula	Elements present	Proportion of atoms
sulphuric acid	H_2SO_4	hydrogen, sulphur, oxygen	1 H : 1 S : 4 O
sodium sulphite	Na_2SO_3		
potassium permanganate	$KMnO_4$		
lithium bromide	LiBr		
sodium arsenate	Na_3AsO_4		
phosphoric acid	H_3PO_4		
strontium chromate	$SrCrO_4$		
potassium hydroxide	KOH		
zinc sulphide	ZnS		
carbon tetrachloride	CCl_4		
iron(III) oxide	Fe_2O_3		
decane	$C_{10}H_{22}$		
silver nitrate	$AgNO_3$		
aluminium chloride	$AlCl_3$		
copper sulphate	$CuSO_4$		

2 Circle the compounds in this list of substances.

barium sulphate	gold	carbon
copper	nitric acid	hydrogen sulphide
iodine	oxygen	mercury
silver chloride	sulphur dioxide	aluminium

Activity 3 **Describing a scientific test** Date:_____

1 Label the diagram.

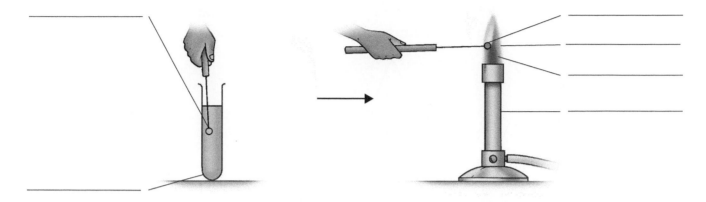

2 Write step-by-step instructions for carrying out a flame test using a solution of copper sulphate.

_____

_____

_____

_____

_____

_____

_____

_____

_____

_____

_____

_____

_____

_____

_____

_____

_____

_____

Activity 4 Identifying elements in compounds

Date: _____

1 Colour each flame to show the result of doing a flame test on a solution containing each of these elements.

lead

copper

lithium

potassium

calcium

sodium

2 Colour each compound the correct colour.

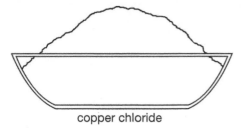
copper chloride

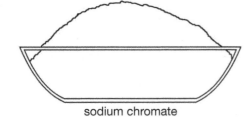
sodium chromate

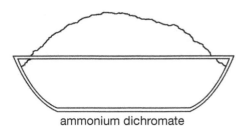
ammonium dichromate

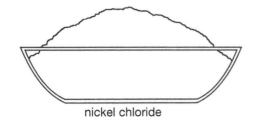
nickel chloride

Activity 5 Test your knowledge

Date:_____

Write a set of clues for this crossword puzzle.

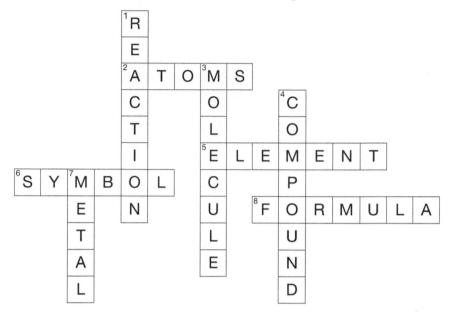

Clues

Down

1 _____

_____

_____

3 _____

_____

_____

4 _____

_____

_____

7 _____

_____

_____

Across

2 _____

_____

_____

5 _____

_____

_____

6 _____

_____

_____

8 _____

_____

_____

More about metals and non-metals

Activity 1 | **Identifying metals**

Date:_____

1 Read the description of each metal carefully. Then complete each information box by writing in the name of the metal.

---

metal:_____

Has been used by humans for thousands of years. It rusts easily so it is mainly used as its alloy (steel), which doesn't rust so easily. The chemical symbol comes from its Latin name, *ferrum*.

---

metal:_____

Light and strong. A very good conductor of heat and electricity. Used to make pots and pans, soft drink cans and foil for wrapping food. Common in the Earth's crust, but it requires lots of energy to refine it. The chemical symbol is Al.

---

metal:_____

This metal is an excellent conductor of heat and electricity. It is orange-brown in colour and it stays shiny because it doesn't corrode easily. When it does corrode it gets a green coating. Its main use is in electrical wires.

---

metal:_____

A heavy metal that is very soft and malleable. When mixed with tin it forms an alloy called solder. Its chemical symbol comes from its Latin name, *plumbum*.

---

metal:_____

A metal that burns with a very bright, white flame. It is light and can be used as an alloy to make 'mag' rims for car tyres. Its chemical symbol is Mg.

---

metal:_____

The only metal that is liquid at room temperature. It is highly poisonous and is sometimes called quicksilver. Often used in thermometers because it expands quickly in response to heat.

---

metal:_____

Iron and steel can be coated with this metal to prevent them from rusting. When mixed with copper, it forms the alloy brass.
Its atomic number is 30.

_____

This is an expensive yellow metal often used to make jewellery. It is fairly unreactive and is found as an element in the Earth's crust. The Latin name for this metal is *aurum*.

metal:_____

A less expensive metal, which is shiny and used in cheaper jewellery. It is more conductive than copper, but it is too expensive to use for wiring.
Its chemical symbol is Ag.

_____

A radioactive metal that is used in nuclear power plants. Atoms of the metal are split at power plants to release large amounts of energy. Its chemical symbol is U.

2 Choose one other metal. Write a description of it here.

metal: _____

Activity 2 **Making alloys** Date:_____

Complete these diagrams to show what metals are mixed to make each alloy.

a)
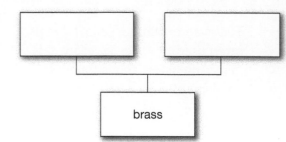

b)
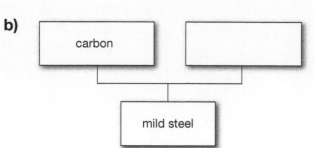

c)
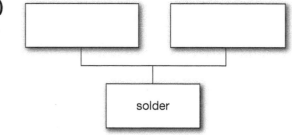

d)
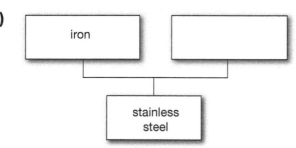

e)
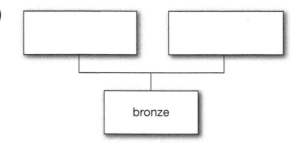

f)
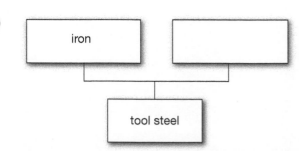

Activity 3 Uses of metals

Date:_____

Complete the third column of this table by filling in the main uses of each metal. You may have to do some research to find the answers.

Metal	Properties	Used to make
aluminium	soft, light, malleable and ductile, corrosion resistant, good conductor, light silver in colour, smooth surface that holds a polish	
copper	excellent conductor, very malleable, ductile, corrosion resistant, can be hardened by working it, rich orange-brown colour	
brass	hard, more brittle than other metals, corrosion resistant, bright golden yellow colour when polished	
tin	soft, malleable, easy to work with, resists corrosion by food, shiny silver colour	
stainless steel	resists corrosion (by food and air), strong, tough, bright and shiny	
mild steel	very ductile, can be shaped easily, corrodes easily	
medium carbon steel	fairly ductile, corrodes easily, easy to shape, can be welded and machined, strong	
high carbon steel	high carbon content reduces ductility and strength, corrodes easily, can be hardened by working	
galvanised steel	fairly ductile, easy to shape and bend, corrosion resistant, dull grey finish	

Activity 4 Gases and their properties   Date:_____

The mind maps show some uses of two important gases.

In each box, write what chemical property of the gas makes it suitable for that purpose.

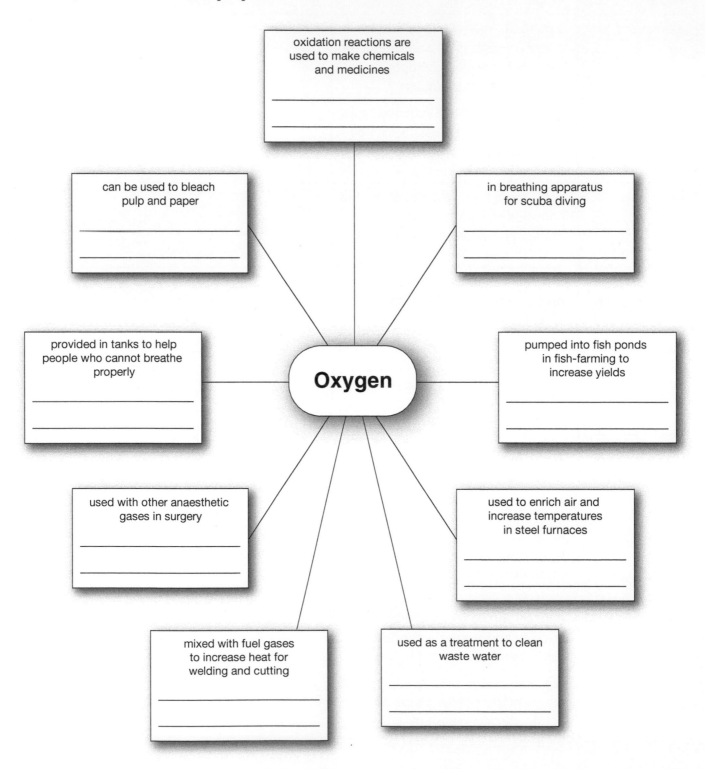

oxidation reactions are used to make chemicals and medicines

_____

_____

can be used to bleach pulp and paper

_____

_____

in breathing apparatus for scuba diving

_____

_____

provided in tanks to help people who cannot breathe properly

_____

_____

Oxygen

pumped into fish ponds in fish-farming to increase yields

_____

_____

used with other anaesthetic gases in surgery

_____

_____

used to enrich air and increase temperatures in steel furnaces

_____

_____

mixed with fuel gases to increase heat for welding and cutting

_____

_____

used as a treatment to clean waste water

_____

_____

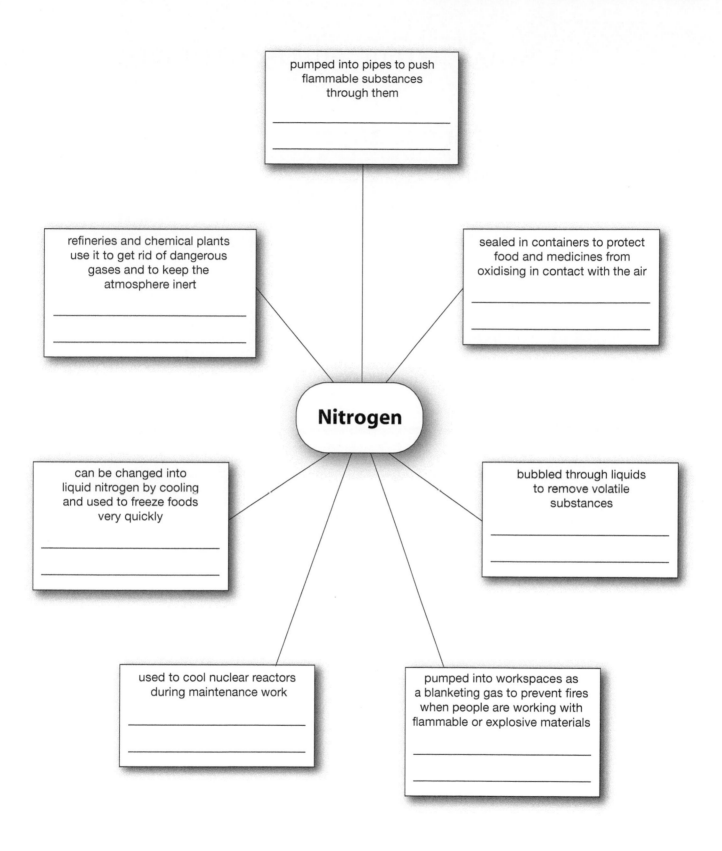

pumped into pipes to push
flammable substances
through them

_____

_____

refineries and chemical plants
use it to get rid of dangerous
gases and to keep the
atmosphere inert

_____

_____

sealed in containers to protect
food and medicines from
oxidising in contact with the air

_____

_____

Nitrogen

can be changed into
liquid nitrogen by cooling
and used to freeze foods
very quickly

_____

_____

bubbled through liquids
to remove volatile
substances

_____

_____

used to cool nuclear reactors
during maintenance work

_____

_____

pumped into workspaces as
a blanketing gas to prevent fires
when people are working with
flammable or explosive materials

_____

_____

Chapter 9　Chemical reactions

Activity 1　Identifying changes

Date:_____

Look at the pictures of changes carefully. Then complete the table.

Picture	What change can you observe?	Is it a physical or chemical change?	Reason for choice

Picture	What change can you observe?	Is it a physical or chemical change?	Reason for choice

Activity 2 | **Writing equations**   Date:_____

The diagrams show some chemical reactions.

Identify and name the reactants and products in each diagram. Then write a word equation underneath to describe each reaction.

a) + O ⟶ Mg O

reactants: _____

products: _____

_____

b) Fe + S ⟶ Fe S

reactants: _____

products: _____

_____

c) S + O O ⟶ O S O

reactants: _____

products: _____

_____

d) 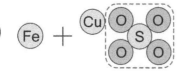 ⟶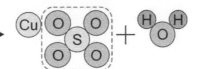

reactants: _____

products: _____

_____

e)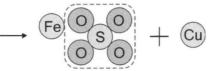

reactants: _____

products: _____

_____

f)  ⟶

reactants: _____

products: _____

_____

g) Ag Cl ⟶ Ag + Cl

reactants: _____

products: _____

_____

h)  ⟶

reactants: _____

products: _____

_____

Activity 3 **Labelling a diagram** Date:_____

James and Sadick heated iron and sulphur together to produce iron sulphide.

1 Label this diagram correctly to show what they did during their experiment.

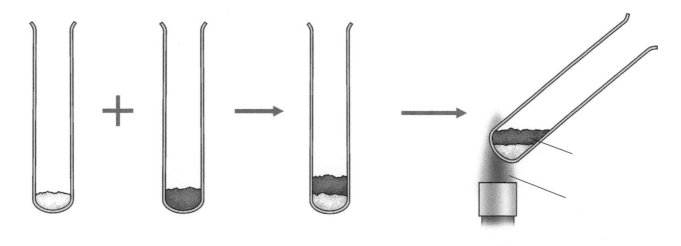

_____ _____ _____

2 Label this diagram to show the reaction they observed.

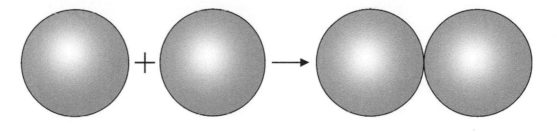

_____ _____ _____

3 Write a word equation to describe the reaction they observed.

_____

Activity 4 **Talking about reactions**   Date:_____

Use the glossary on pages 155 to 158 of your coursebook, and your own knowledge, to find the meaning of each of these key words from this chapter.

Use each word in a sentence to show what it means.

Key words	Sentence
physical change	
chemical reaction	
reactant	
product	
combustion	
combination	
decomposition	

 **Chapter 10** Rocks and weathering

Observing and comparing rocks

Date:_____

You will need

- three different pieces of rock • a magnifying glass • some water

Method

Observe your rock samples carefully – how are they similar, and how are they different?

Questions

1 Complete this table to record the similarities and differences.

Characteristics	Sample 1	Sample 2	Sample 3
name of rock			
colour			
shape			
texture			
mass			
other			

2 Use a magnifying glass to look closely at your rock samples. Wet your rock with a few drops of water and look again. On separate paper, draw and label what you saw under the magnifying glass.

Activity 2 **The rock cycle**   Date:_____

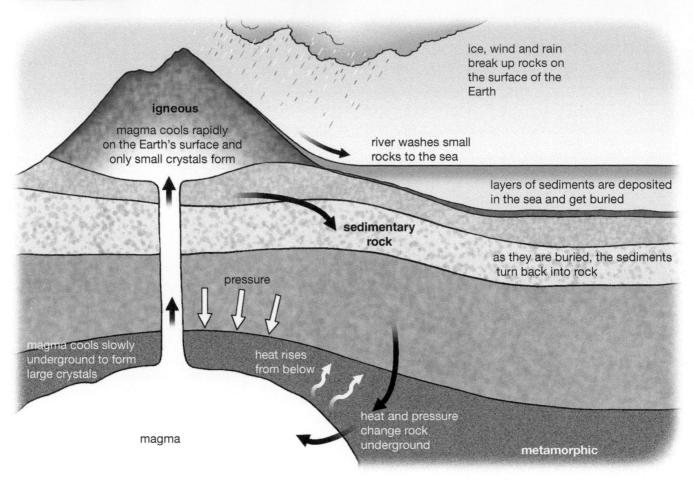

igneous

magma cools rapidly
on the Earth's surface and
only small crystals form

ice, wind and rain
break up rocks on
the surface of the
Earth

river washes small
rocks to the sea

layers of sediments are deposited
in the sea and get buried

sedimentary
rock

as they are buried, the sediments
turn back into rock

pressure

magma cools slowly
underground to form
large crystals

heat rises
from below

heat and pressure
change rock
underground

metamorphic

magma

1 What happens to rocks on the surface of the Earth?

_____

_____

2 How does rock from the surface of the Earth get back underground?

_____

_____

3 Which two processes work underground to form and change rocks?

_____

_____

4 What affects the size of particles in igneous rocks?

_____

_____

Activity 3 Describing an experiment Date:_____

A teacher did this experiment to show her class what happens to a piece of rock when it is heated and cooled. Look at the diagrams carefully.

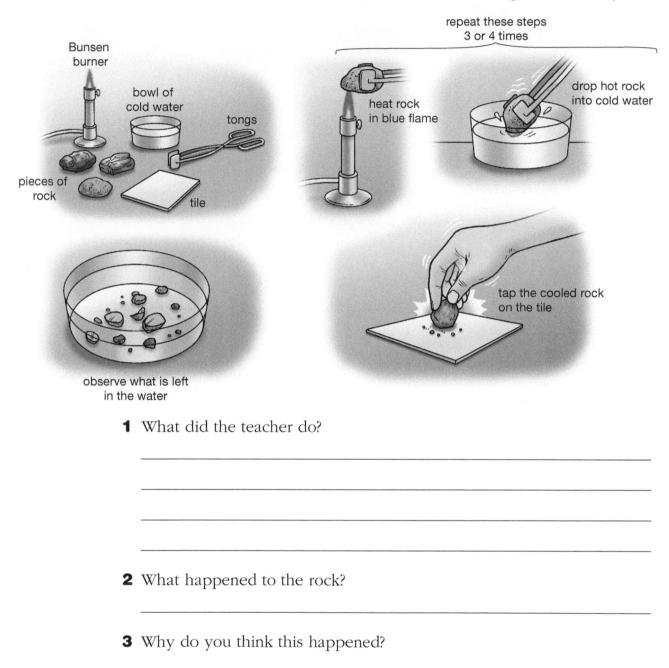

repeat these steps
3 or 4 times

Bunsen burner

bowl of cold water

tongs

pieces of rock

tile

heat rock in blue flame

drop hot rock into cold water

observe what is left in the water

tap the cooled rock on the tile

1 What did the teacher do?

_____

_____

_____

_____

2 What happened to the rock?

_____

3 Why do you think this happened?

_____

_____

4 The teacher heated the rock more than rock would be heated naturally on Earth. Why do you think she needed to do this?

_____

_____

Activity 4 Conducting an investigation   Date:_____

> How does the strength of the acid affect the rate at which limestone dissolves?

> I wonder if the amount of time that the limestone is exposed to the acid makes any difference?

> Do you think the size of the rock affects the rate at which it dissolves?

Choose *one* question from those above.

My question:

_____

_____

Rewrite the question as a hypothesis.

My hypothesis:

_____

_____

Design and carry out a fair test to test the validity of the hypothesis you have developed.

Record your planning and results here.

I will test:

_____

_____

I will need:

_____

_____

I will measure:

_____

_____

I will change these things:

_____

_____

I will keep these things the same:

_____

_____

I will record my data like this:

I learned:

_____

_____

Activity 5 **Test your knowledge** Date:_____

The following statements are *false*.

Rewrite each statement to make it *true*.

1 The Earth's crust is made up of only one type of rock.

_____

_____

2 All rocks are formed deep below the Earth's surface.

_____

_____

3 Once formed, a rock will last forever.

_____

4 Rocks are classified according to their colour and size.

_____

_____

5 Erosion is the process by which rocks are broken down.

_____

_____

6 Rocks can be heated and cooled without anything happening to them.

_____

_____

7 Limestone is a hard igneous rock.

_____

8 Granite will dissolve if you put it in vinegar.

_____

9 Coal is an example of a metamorphic rock.

_____

10 It is easy to make coal. If you put dead plants in the ground you will get coal in a short time.

_____

_____

Chapter 11 Magnetism

Activity 1   **Magnetic forces**   Date: _____

1 The magnets below are attracted to each other. Write the correct
pole – N or S – on each unlabelled magnet end.

2 The magnets below are repelled by each other. Fill in the correct
poles.

3 The arrows on this diagram show you whether the magnets are being
attracted or repelled by each other. Fill in the correct poles.

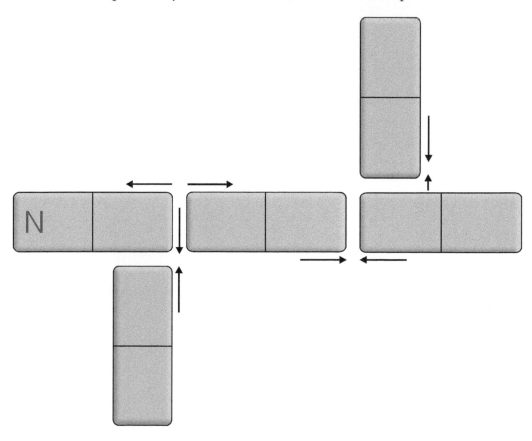

Activity 2 **Different field patterns** Date:_____

1 Arrange two magnets as shown in each of these diagrams.

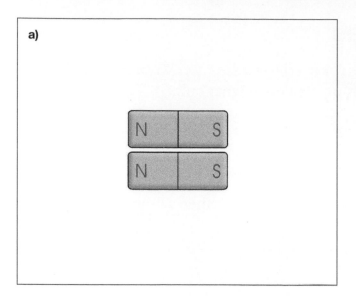

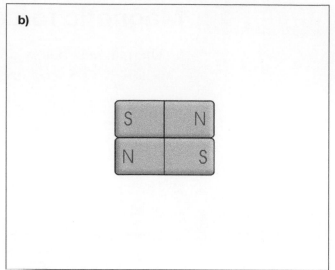

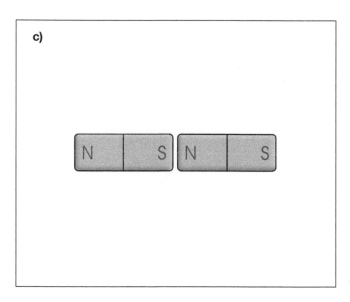

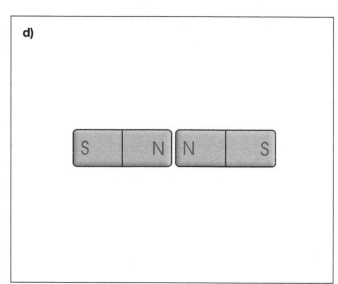

2 Cover the magnets with a sheet of paper.

3 Sprinkle iron filings onto the paper to see the shape of the magnetic field in each case.

4 Draw field lines on each diagram to show your results.

Activity 3 Electromagnets at work   Date:_____

The diagram shows you how an electric door bell works. Study the
diagram carefully and fill in the missing words in the passage below.
The passage continues on the next page.

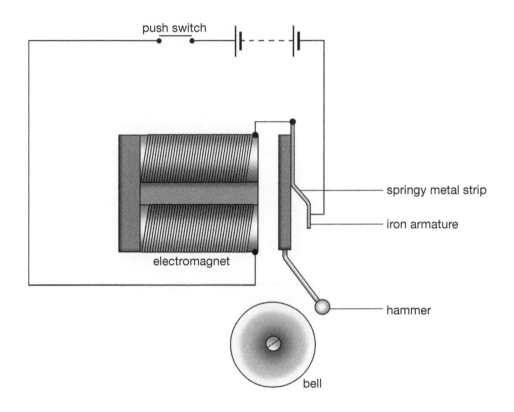

When you press the _____ the circuit is closed (completed)

and a current passes through the _____. This turns the

coiled wire into an _____.

The electromagnet exerts a force of attraction on the iron

_____. This _____ the iron armature towards

the electromagnet and the _____ hits the metal dome of the

bell.

When the iron armature moves towards the electromagnet it pulls away

from the _____ and a break develops in the circuit.

When the circuit is broken, the _____ loses its magnetism

and the iron _____ moves back to its original position. This

causes the circuit to _____ again and the _____

starts to exert a force on the iron armature once again.

This continues, with the _____ hitting the _____

over and over again as the circuit is _____

and _____.

When you stop pushing the _____, the bell

_____ ringing.

 **Investigating magnets**   Date:_____

Find ten examples of magnets used in your local environment.

Complete this table to record what you found out.

Place where magnet is used	Type of magnet used	What the magnet does

Activity 5 **Using scientific words**   Date:_____

Use your scientific knowledge and your dictionary to find the meaning of each of these key words from this chapter.

Use each word in a sentence to show what it means.

1 magnet

_____

_____

2 electromagnet

_____

_____

3 magnetic field

_____

_____

4 magnetic

_____

_____

5 non-magnetic

_____

_____

6 poles

_____

_____

7 attraction

_____

_____

8 repulsion

_____

_____

Chapter 12 Light

Activity 1 **Using sunlight to tell the time**

Date:_____

Here is a diagram of a sundial.

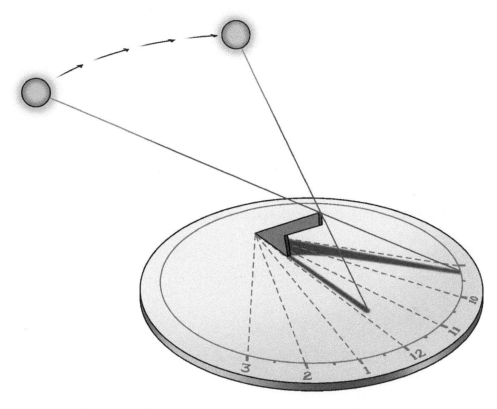

1 Write labels on the diagram to explain how the sundial works.

2 When you set up a sundial, you have to do two things:
- make sure that the base is level
- make sure that the piece that makes the shadow (the gnomon) points south in the southern hemisphere and north in the northern hemisphere.

Explain why these two things are important.

_____

_____

_____

_____

Activity 2 Drawing angles of reflection Date:_____

Use a ruler and a protractor to draw in a reflected ray from the mirror in each of these diagrams. In some diagrams, you may have to find the normal first.

a)

b)

c)

d)

e)

f)

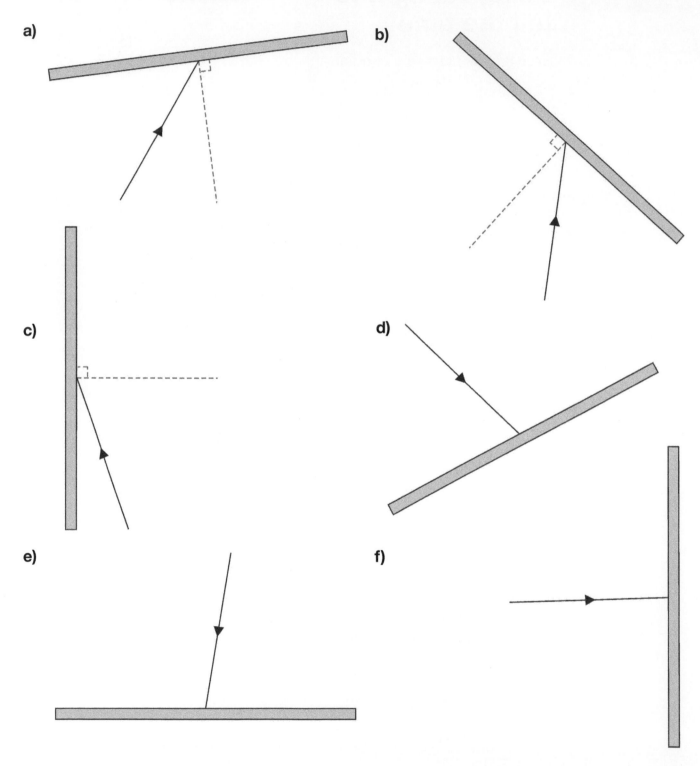

Activity 3   **Follow the light**

Date:_____

The diagram shows a beam of light from a strong light source.

The beam is reflected by a series of flat mirrors.

Draw the path the light takes from point A to point B.

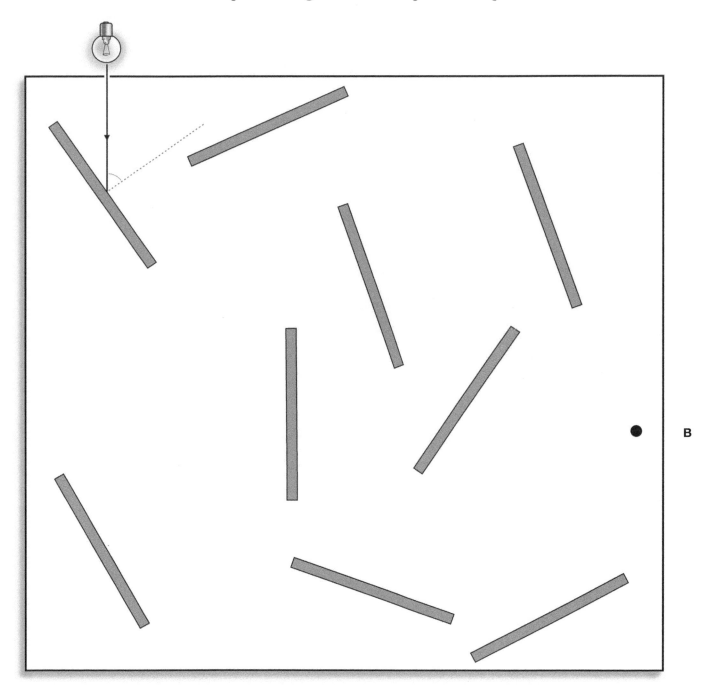

B

Activity 4 **Investigating images**   Date:_____

You are going to investigate images formed in a mirror.

You will need two flat mirrors.

1 Look at your reflection in the mirror. What can you say about it?

_____

_____

   a) What happens to your image when you move the mirror towards you or away from you?

   _____

   _____

   b) Is your image identical to you?

   _____

   c) Is your image upside down or the right way up?

   _____

2 Print your name in capital letters and look at it in the mirror.
   a) Is it easy to read? _____
   b) Which letters look the same in the mirror as on the paper?

   _____

   c) Which letters look different in the mirror and on the paper?

   _____

   d) Why do you think this happens?

   _____

   _____

3 Hold two mirrors at 90° to each other.
   a) How many reflections do you think you will see if you put an object between the two mirrors? _____
   b) Put an object between the mirrors. How many reflections can you see? _____
   c) Was your prediction correct? _____
   d) Investigate what happens if you change the angle between the two mirrors. Record your observations in your notebook.

Activity 5 **Mixing colours**   Date:_____

You will need

- a paper cup ● compasses ● a pencil ● scissors ● white card
- coloured pens or pencils

Method

Use compasses to draw a 15 cm diameter circle on a piece of white card, and cut it out. Divide it into sections like this and colour the sections correctly. Try to colour smoothly and evenly.

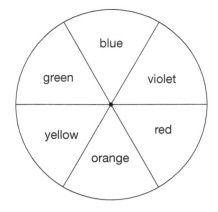

Make a hole in the centre of your colour wheel and push the pencil through it.

Make a hole in the bottom of the cup, slightly bigger than the pencil.

Put the pencil and colour wheel together as shown. The point of the pencil should rest on the desk.

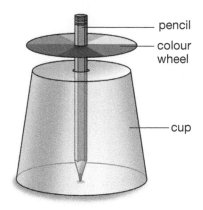

Spin the pencil quickly and watch what happens to the colour wheel.

Keep trying till the colours on the disk blend together and you can only see white.

Explain why this happens.

_____

_____

_____

Sound and hearing

How musical instruments make sounds

Date:_____

Next to each instrument, write down how you can make it produce vibrations and make sounds.

drum
how it makes sound:

guitar
how it makes sound:

xylophone
how it makes sound:

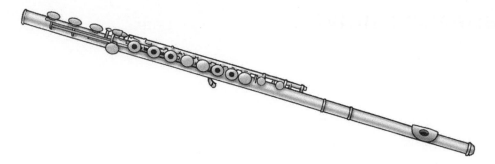

flute
how it makes sound:

maraca
how it makes sound:

saxophone
how it makes sound:

Activity 2 **Investigating how sound travels**

Date:_____

You will need

- a two-litre plastic bottle • a plastic bag • an elastic band
- scissors • a tea light candle

Method

Following the instructions below, carry out an experiment to show that sound vibrations travel through the air.

Be very careful with the matches and the candle flame!

Step 1
Cut the bottom off the bottle. Cut a piece of plastic big enough to cover the end of the bottle.

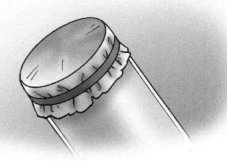

Step 2
Stretch the piece of plastic tightly over the end of the bottle. Hold it in place with the elastic band.

Step 3
Light the candle. Hold the bottle so that the neck is about 2.5 cm away from the candle flame.

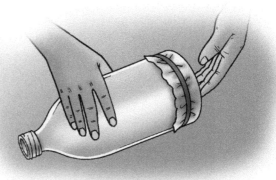

Step 4
Hold the bottle still. Tap the piece of plastic sharply with your fingertips.

Now answer the questions on the next page.

Questions

1 What happens to the candle flame?

_____

_____

2 Label this diagram to show what happened in your experiment.

Activity 3 Sound waves

Date:_____

1 Describe the sound shown by each oscilloscope pattern. Use the words in this box in your descriptions.

> loud   soft   high-pitched   low-pitched

a)

_____

b)

_____

c)

_____

d)

_____

e)

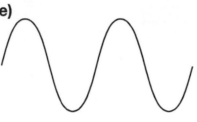

_____

f)

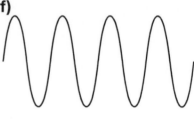

_____

2 The diagram on the first grid below shows the oscilloscope pattern made by a musician hitting a note on a xylophone.

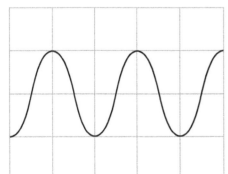 

On the two blank grids, draw the wave patterns you would see if the musician played:

a) the same note louder

b) a lower-pitched note at the same loudness.

Activity 4 **Human hearing** Date:_____

1 Label each part in the diagram of the human ear with its correct name.

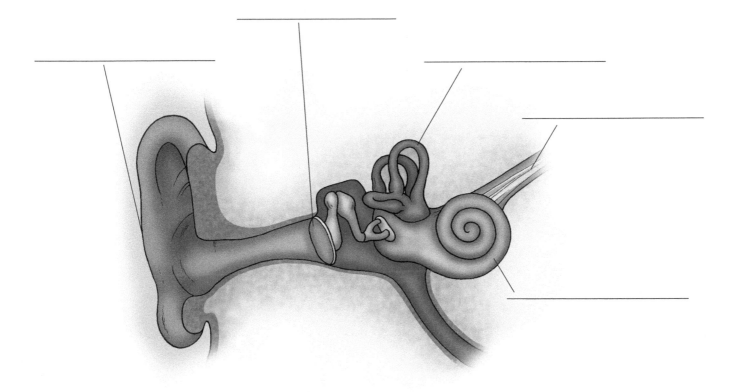

2 Draw a flow diagram to show how we hear.

3 Write down *two* ways in which humans can make sounds.

_____

_____

Chapter 14 | Energy transformations

Activity 1 | Identifying forms of energy Date:_____

1 Give *two* examples of each of these different forms of energy that you experience in your daily life.

a) stored energy

_____   _____

b) sound energy

_____   _____

c) heat energy

_____   _____

d) kinetic energy

_____   _____

e) light energy

_____   _____

f) electrical energy

_____   _____

2 Look at the sources of energy listed below. For each one, give *two* forms of energy that it can be changed into to make it more useful to us.

a) a lump of coal

_____   _____

b) a barrel of oil

_____   _____

c) sunlight

_____   _____

d) wind

_____   _____

e) a piece of wood

_____   _____

Activity 2 Energy transformations   Date:_____

1 Complete this table.

Input	Transducer	Output
electrical energy from battery	torch	
electrical energy from mains	stove	
electrical energy from battery	door bell	
chemical energy from petrol	lawn mower	
chemical energy from gas	barbecue grill	
kinetic energy from wind	windmill	
chemical energy from food in body	bicycle	

2 Choose *three* of these transformations and draw an energy transfer diagram for each.

Activity 3 Sankey diagrams

Date:_____

1 Read the information in the box carefully.

> Very little of the energy from the Sun that reaches the Earth's atmosphere reaches the Earth itself. 30% of the energy is reflected back into space before it even gets through the atmosphere and 47% is absorbed as it passes through the atmosphere. 23% of the energy is used by the water cycle, but less than 1% is used to produce winds and ocean currents. Only about 0.02% of the Sun's energy is used by plants in photosynthesis.

2 Now label the arrows on the Sankey diagram correctly.

Activity 4 Building a turbine

Date:_____

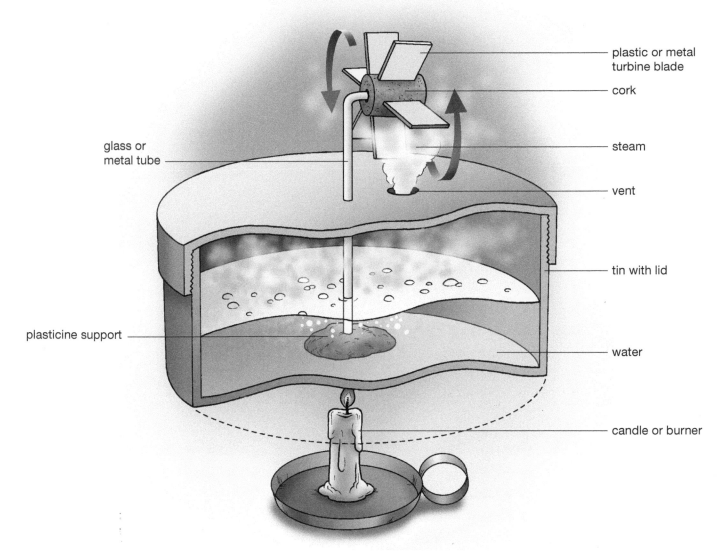

plastic or metal
turbine blade

cork

steam

glass or
metal tube

vent

tin with lid

plasticine support

water

candle or burner

1 Find the materials you need and build a steam turbine like the one in the picture.

2 Test your turbine to see if it works.

3 Think about how you could improve your turbine.

4 On separate paper, write a report showing how you built the turbine. Use these headings to organise your report.

Aim	**Materials used**	**Method**
How we tested the turbine		**Results of the test**
Evaluation of turbine		**Suggestions for improving the turbine**

Activity 5   Generating electricity

Date:_____

The flow diagram shows how electricity is generated in a coal-burning power station and then transmitted to the national grid.

Label the diagram correctly.

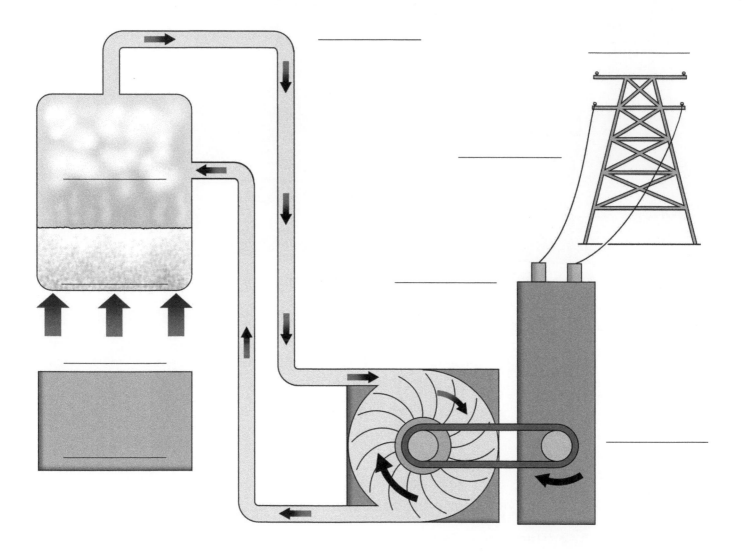

Speed, time and distance

Activity 1 **Converting speeds**   Date:_____

The speeds down the left of this scale are given in kilometres per hour.

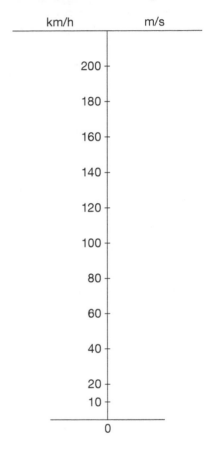

Fill in the equivalent speeds in metres per second on the right-hand side of the scale.

Remember:

- to change kilometres to metres, multiply by 1000
- this will give you metres per hour
- to change metres per hour to metres per minute, divide by 60
- to change metres per minute to metres per second, divide by 60 again.

Activity 2 **Measuring the speed of cars**

Date:_____

Aim

To find the speed travelled by different cars over 100 metres of road.

You will need

- a stopwatch
- a safe stretch of road

Method

Work in pairs.

Mark off 100 metres on the side of the road.

Stand at the starting point.

Take turns to measure the time it takes each car to cover 100 metres. Record the data in this table.

Type of car	Time taken to cover 100 metres (in seconds)	Speed in km/h

Measure the times for ten cars.

Work out the speed of each car in km/h
(speed in km/h = 360 ÷ time taken to travel 100 m).

Calculate the average speed of cars on that stretch of road in km/h.

Activity 3 Drawing a graph

Date:_____

Use the data from Activity 2 to draw a graph showing the speed travelled by three different cars.

Give your graph a title. Label each line with the type of car.

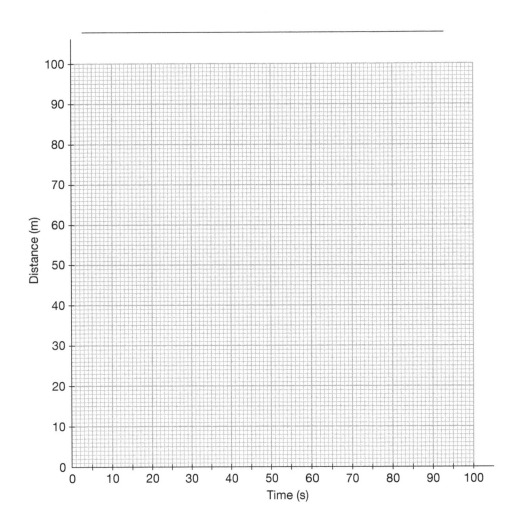

Activity 4 Answering questions about a graph

Date:_____

A group of pupils went on a weekend hike.

They drew this graph of their journey.

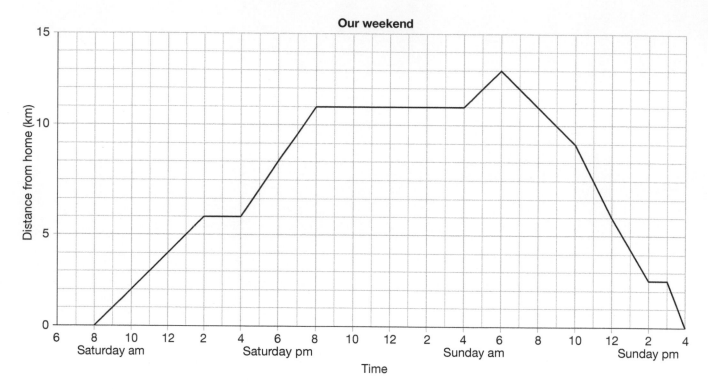

1 At what time did the hike start?

_____

2 When did the students return home?

_____

3 What was the furthest distance they travelled from home?

_____

4 How many times did they rest during the day on Saturday?

_____

5 What was their speed from 8 am to 2 pm on Saturday?

_____

_____

6 When did the students walk the fastest? How do you know this?

_____

_____

7 How far did the students walk altogether on their hike?

_____